Mojisola Ojebode

Essential Oils in Food Pathogen Mitigation

AF 153187

Mojisola Ojebode

Essential Oils in Food Pathogen Mitigation

LAP LAMBERT Academic Publishing

Impressum / Imprint

Bibliografische Information der Deutschen Nationalbibliothek: Die Deutsche Nationalbibliothek verzeichnet diese Publikation in der Deutschen Nationalbibliografie; detaillierte bibliografische Daten sind im Internet über http://dnb.d-nb.de abrufbar.

Alle in diesem Buch genannten Marken und Produktnamen unterliegen warenzeichen-, marken- oder patentrechtlichem Schutz bzw. sind Warenzeichen oder eingetragene Warenzeichen der jeweiligen Inhaber. Die Wiedergabe von Marken, Produktnamen, Gebrauchsnamen, Handelsnamen, Warenbezeichnungen u.s.w. in diesem Werk berechtigt auch ohne besondere Kennzeichnung nicht zu der Annahme, dass solche Namen im Sinne der Warenzeichen- und Markenschutzgesetzgebung als frei zu betrachten wären und daher von jedermann benutzt werden dürften.

Bibliographic information published by the Deutsche Nationalbibliothek: The Deutsche Nationalbibliothek lists this publication in the Deutsche Nationalbibliografie; detailed bibliographic data are available in the Internet at http://dnb.d-nb.de.

Any brand names and product names mentioned in this book are subject to trademark, brand or patent protection and are trademarks or registered trademarks of their respective holders. The use of brand names, product names, common names, trade names, product descriptions etc. even without a particular marking in this work is in no way to be construed to mean that such names may be regarded as unrestricted in respect of trademark and brand protection legislation and could thus be used by anyone.

Coverbild / Cover image: www.ingimage.com

Verlag / Publisher:
LAP LAMBERT Academic Publishing
ist ein Imprint der / is a trademark of
OmniScriptum GmbH & Co. KG
Bahnhofstraße 28, 66111 Saarbrücken, Deutschland / Germany
Email: info@lap-publishing.com

Herstellung: siehe letzte Seite /
Printed at: see last page
ISBN: 978-3-659-83045-7

Table of contents

Chapter Three

Chapter Four

Chapter Five

List of Tables

List of Figures

Introduction

Food safety is an increasingly important public health issue despite continuous improvements in food production techniques and hygiene [1]. In industrialized countries, about 30% of people suffer from a food borne disease each year [1]. Microbial activity is a primary mode of deterioration of foods and many microorganisms are responsible for such occurrences that contribute to loss of food quality and safety. Antimicrobial chemicals had been largely used as foods preservatives. However, but the uncontrolled concentrations applied increase the risk of toxic residues in such food products [2]. Hence, the need for new methods of reducing or eliminating food borne pathogens, possibly in combination with existing methods of food processing techniques [3]. More so, Western society seems to be experiencing an increasing trend in the demand for 'green' solutions, desiring fewer synthetic food additives and products [3]. This is in the hope of limiting the adverse effects of synthetic chemicals on health and the environment. In addition, the World Health Organization has called for a worldwide reduction in salt consumption to reduce the incidence of cardio-vascular disease [4]. Salt has often been used as a means of preserving processed food; hence, a reduction in its level will necessitate the use of other additives to maintain the safety of food. One such possibility is the use of essential oils for food pathogen mitigation.

Essential oils (EOs) are aromatic, volatile, oily and coloured liquids, obtained from plant parts (flowers, buds, seeds, leaves, twigs, bark, herbs, wood, fruits and roots). They are soluble in lipids and organic solvents that have a lower density than water. Essential oils are generally stored by the plant in secretory cells, cavities, canals or epidemic cells [5]. Essential oils contain numerous secondary metabolites that can slow down or inhibit the growth of bacteria and other pathogens. They are commonly extracted using the method of steam or water distillation for commercial production as well as empyreumatic distillation. Essential oils or their components have been shown to exhibit antimicrobial, antiviral, anti-mycotic, anti-toxigenic, antiparasitic and insecticidal properties [3][6][7]. When essential oils are used against pathogens, they target particularly the membrane and cytoplasm, and in some cases, they affect cell morphology [5]. Essential oils as pathogenic agents are recognized as user- and environmentally- friendly natural substances and they have been considered at low risk for resistance development by pathogenic microorganisms [2]. This book presents a review of the main components of essential oils extracted from edible herbs and spices, their mechanisms of action and effects against food-borne pathogens which could be harnessed for food preservation. It also provides an experimental study of the antibacterial activity of the essential oils of lemon grass (*Cymbopogon citratus*) a medicinal and insecticidal plant and orange peels (*Citrus sinensis*) which could have been trashed as case study.

Chapter One

1.1 Food Pathogens

Food pathogens are infectious agents such as viruses, bacteria or parasites that contaminate food items and cause diseases or illnesses when such items are consumed.

1.1.1 Bacteria food pathogens

Bacteria can be classified into two classes, these are, Gram-positive and Gram-negative bacteria.

1.1.2 Gram-positive bacteria

Gram-positive bacteria in food include *Listeria monocytogenes, Clostridium botulinum, Bacillus cereus, Staphylococcus aureus* and *Clostridium perfringens*. The cell wall of Gram-positive bacteria consists of approximately 90%-95% peptidoglycan, to which other molecules, such as proteins are linked (Figure 1) [5].

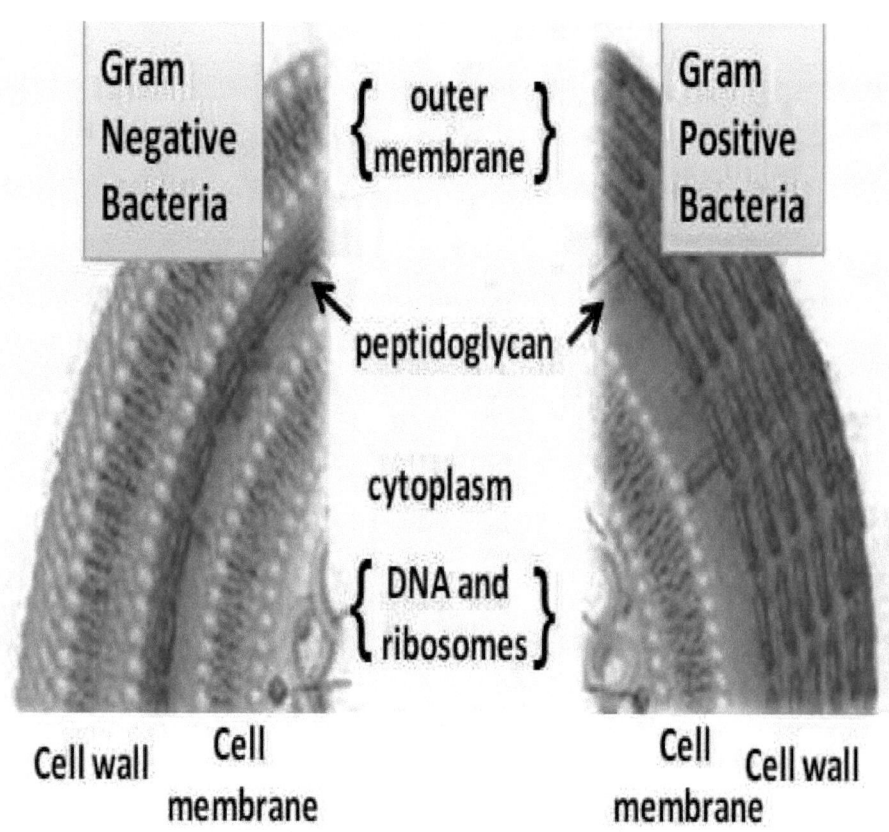

Figure 1. Schematic representation of the envelopes of Gram-positive and Gram-negative bacteria [5].

Examples of gram-positive bacteria in food.

- *Staphylococcus aureus (S. aureus)*

Staphylococcus aureus is often found on the skin, throat and nostril of many healthy people and animals. Heating is not sufficient to get rid of this microorganism. It causes illness when transmitted to food products where it can multiply and produce harmful toxins. *S. aureus* could be present on food items such as meat and meat products, unpasteurized milk and dairy products, custard or cream-filled pastries, egg foods, cheese, prawn and salad containing potatoes and pasta.

Staphylococcus aureus is responsible for a range of illnesses from impetigo, furuncles, scalded skin syndrome, to pneumonia, meningitis, endocarditis and sepsis. The symptoms include, vomiting, diarrhoea, abdominal cramps, death (rare). The infective dose is 10^5-10^6 bacteria/ml and is often transmitted to food by infected food handlers.

- *Listeria monocytogenes*

Listeria monocytogenes is capable of growth at refrigeration temperature (5^0C). It is salt tolerant (halophile) and has an incubation period of 3-70 days. Three types of diseases are caused by this bacteria. These are:

- Adult listeriosis

The symptoms of this disease include fever, muscle/ headaches, nausea or diarrhoea, meningitis, encephalitis, bacteremia. If the nervous system is affected, stiff neck, confusion, loss of balance or convulsion and even death may result.

- Listeria gastroenteritis- Febrile gastroenteristis

In normal host causes fever, diarrhoea, vomiting but no death attached.

- Listeriosis in pregnancy/neonatan listeriosis

This could be acquired transplacentally or during or after delivery, listeriosis of a new born, neonatal sepsis/meningitis, still births, spontaneous abortion, death.

Food items contaminated by this bacteria include unpasteurized milk and fruit juices, soft cheese, raw meat, ice cream, ready-to-eat food and meat, vegetable.

- *Bacillus cereus*

  B. cereus is a spore-forming microorganism with infective dose minimum of 10^5/g. It is responsible for the following diarrheal diseases; watery diarrhoea, abdominal cramps and pain. Emetic diseases such as nausea and vomiting; heat-stale enterotoxin produced by growing cells in the food; death in rare cases.

  Food items contaminated by *Bacillus cereus* causing diarrhoea include meat, milk, vegetables and fish while those that cause emetics include rice-products, starchy foods, casseroles, puddings, soups.

- *Clostridium botulinum*

  Clostridium botulinum is an obligate anaerobe. The heat-resistant spores can survive at 100^0C for several hours, but spores are killed at 120^0C for 30 minutes. Botulinum neurotoxin is easily destroyed by heat after cooking for 30 minutes at 80^0C. Botulism occurs in three forms:

- Foodborne botulism

  This causes weariness, weakness, dizziness, double vision, difficult swallowing and speaking, paralysis and death which can be caused by the toxin produced when contaminated food is eaten.

- Wound and infant botulism

  In this case, neurotoxin is produced in infected tissues and in the gastrointestinal tract and death may occur.

Food contaminated include home-made canes, meat products (sausages, black pudding), baked potatoes, garlic/oil mixtures, low acid canned foods, honey (infant botulism).

- *Clostridium perfringens*

Clostridium perfringens is a spore forming bacteria and an obligate anaerobe. Infection is caused by sporulation of bacterial cell in the intestine. Symptoms include abdominal pain, nausea, diarrhoea, fever, headache and death in a few cases. Contaminated foods include stews, gravies, beans, meat, poultry and fish.

1.1.3 Gram negative bacteria

- *Campylobacter jejuni*

Campylobacter jejuni is a microaerophilic bacteria that is sensitive to heat, drying, acidic conditions and disinfectants. Campylobacteriosis is the most commonly identified bacterial cause of diarrheal illness in the world. Symptoms of infection include watery diarrhoea, vomiting, headache, fever, muscle-pain and sometimes death. Children are at an increased risk of infection. Foods contaminated by *Campylobacter jejuni* are poultry and dairy products.

- *Salmonella enteriditis*

Salmonella enteriditis was named after Daniel Elmer Salmon, an American veterinary scientist. It is an enteric bacteria and all strains are pathogenic colonizing intestinal epithelium. The bacteria is responsible for two diseases namely:

- Enterocolitis salmonellosis: (most commonly by *Staphylococcus enterica serovar typhimurium*). Symptoms include headache, chills, vomiting, diarrhoea, fever and rarely death.

- Typhoid fever

Foods contaminated by *Salmonella enteriditis* include; poultry, meat, egg and egg products, sliced melons, vegetables, chocolates.

- *Escherichia coli*

E. coli colonizes the small intestine and produces verotoxins. Symptoms of *E. coli* infection include dehydration, bloody diarrhoea, nausea, vomiting, kidney failure and destruction of red blood vessels. Contaminated foods are beef, yogurt, lettuce, contaminated fruits and vegetables, unpasteurized milk and juice.

- *Shigella dysenteria*

Genus is named after Kiyoshi Shiga. Infection is called Shigellosis or bacillary dysentery. Symptoms of infection include abdominal pain, diarrhoea, fever, blood in faeces, nausea, dehydration and vomiting. *Shigella* toxin (verotoxin) causes severe diarrhoea, massive tissue inflammation and destruction and rarely death. Foods contaminated by *Shigella dysenteria* are salads and seafoods.

- *Yersinia enterocolitica*

Yersinia enterocolitica is capable of growth at 2^0C. Symptoms of infection are severe abdominal pain, diarrhoea, fever, vomiting, and arthritis in adults, pseudoappendiatis and rarely death. Contaminated foods include raw pork or beef, drinking water and milk products.

1.2 Both gram-negative and gram-positive microorganisms are susceptible to the effects of essential oils

Generally, essential oils are slightly more active against gram-positive than gram-negative bacteria based on the studies done to investigate the action of whole EOs against food spoilage organisms and foodborne pathogens [3]. However, the presence of an outer membrane surrounding the cell wall restricts diffusion of hydrophobic compounds through the lipopolysaccharide covering of gram-negative organisms makes them less susceptible to the action of antimicrobial agents. Nevertheless, not all studies have come to the conclusion that gram-positive bacteria are more susceptible to EOs. According to Wan *et al*, 1998, *A. hydrophila* which is a gram-negative bacteria appears to be one of the most sensitive species. In a study, mint (*Mentha piperita*) essential oil showed a greater reduction in the viable count of *Salmonella enteritidis* than for *Listeria monocytogenes* when added to the Greek appetisers tzatziki and taramosalata [8]. In another study, there was no obvious difference measured between the susceptibility of gram-positives and gram-negatives after 24 hours, but the inhibitory effect was more often extended to 48 hours with gram-negative than gram-positive organisms [9]. Dorman and Deans, 2000 carried out a study to verify a previous study testing 50 commercially available EOs against 25 genera. In contrast to the previous study which found no evidence for a difference in sensitivity between gram-positive and gram-negative organisms, they revealed that gram-positive bacteria were indeed more susceptible to two of the freshly distilled EOs tested and equally sensitive to four other EOs than were gram-negative species. It was postulated that the different components of EOs exhibit varying degrees of activity against gram-positive and gram-negative bacteria [10].

Chapter Two

2.1 Essential oils

Essential oils are concentrated natural products produced by aromatic plants as secondary metabolites. They possess strong smells and are coloured. Essential oils are so named because the oils form the 'essence' of fragrance of the plants from which they are derived. Hence, the essential in this case is not the same as essential amino acid which is used to classify amino acids that cannot be synthesized by the body but can be obtained from diet. The oils are present as variable mixtures of majorly terpenoids, especially, mono- (C_{10}), sesqui- (C_{15}) and di- (C_{20}) terpenes may also be present. Other molecules include acids, alcohols, aldehydes, aliphatic hydrocarbons, acyclic esters or lactones and rarely, nitrogen and sulphur containing compounds, coumarins and homologues of phenylpropanoids [5]. Essential oils can be present in all parts of aromatic plants, these include, buds, leaves, flowers, seeds, twigs, stems, fruits, roots, wood or bark. General storage points in plants include secretory cells, canals, cavities, glandular trichomes or epidermic cells. EOs are extracted for their numerous activities which include anti-oxidant, anti-fungal, anti-bacterial, anti-mutagenic, antiseptic, anti-inflammatory, anti-diarrhoeal and are often used as fragrances, in food preservation and as analgesics, sedatives, anti-inflammatories, spasmolytics and local anaesthetics [11].

A number of plant derived essential oils have been screened for antimicrobial activities (Figure 1). Cowan (1999) reported that approximately 3,000 essential oils are currently known so far [12].

Figure 2. Some important medicinal plants widely used for the extraction of essential oils.

A) Cinnamon (*Cinnamomum zeylancium*); B) Thyme (*Thymus broussonetii*); C) Rosmery (*Rosmarinus officinalis*) D) Oregano (*Origanum vulgare*); E) Clove (*Syzygium aromaticum*); F) Worm wood (*Artemisa arbrescens*); G) Caraway (*Carum carvic*); H) Lemon grass (*Cymbopogon citratus*); I) Sage (*Salvia officinalis*) [13].

2.2 Composition of essential oils

The main components of essential oils are terpenes, terpenoids, and phenylpropenes. The phenolic components are chiefly responsible for the anti-bacterial properties of EOs [14]. The major components of some EOs with antibacterial activity are presented in Table 1 and the structural formula of a number of antibacterial components are presented in Table 1.

Table 1

Major components of selected [a] Essential Oils that exhibit antibacterial properties.

Common name of Essential oils	Latin name of plant source	Major components	Approximate % composition	References
Cilantro	*Coriandrum sativum*	Linalool E-2-decanal	26% 20%	(Delaquis *et al.*, 2002)
Coriander	*Coriandrum sativum* (seeds)	Linalool E-2-decanal	70% -	(Delaquis *et al.*, 2002)
Cinnamon	*Cinnamomum zeylandicum*	Trans-cinnamaldehyde	65%	(Lens-Lisbonne *et al.*, 1987)
Oregano	*Origanum vulgare*	Carvacrol Thymol γ -Terpinene	Trace-80% Trace-64% 2-52%	(Lawrence, 1984) (Charai *et al.*, 1996) (Kokkini *et al.*, 1997)
Rosemary	*Rosmarinus officinalis*	Apinene Bornyl acetate Camphor 1,8-cineole	2-25% 0-17% 2-14% 3-89%	(Daferera *et al.*, 2000, 2003; Pintore *et al.*, 2002)
Clove (bud)	*Syzygium aromaticum*	Eugenol Eugenyl acetate	75-85% 8-15%	(Bauer *et al.*, 2001)
Thyme	*Thymus vulgaris*	Thymol Carvacrol γ-Terpine p-Cypene	10-64% 2-11% 2-31% 10-56%	(McGimpsey *et al.*, 1994; Lens-Lisbonne *et al.*, 1987.

[a]Essential Oils which have been shown to exert antibacterial properties in vitro or in food models [3].

2.2.1 Terpenes

Terpenes are hydrocarbons synthesized within the cytoplasm of vegetal cells and are formed from the combination of several isoprene units (C_5H_8). Their synthesis occurs in the mevalonic acid pathway starting from acetyl CoA. Cyclases are involved in rearranging the hydrocarbon backbone into a cyclic structure [15]. Monoterpenes ($C_{10}H_{16}$) and sesquiterpenes ($C_{15}H_{24}$) are the most common terpenes, longer chain terpenes such as diterpenes ($C_{20}H_{32}$) and triterpenes ($C_{30}H_{40}$) are also present in the plant cell. Limonene, p-Cymene, terpinene, sabinene and pinene are the most known terpenes. Most terpenes do not show high antimicrobial activity. In-vitro tests indicate that terpenes show little or no effective antimicrobial activity when used as singular compounds.

Examples of terpenes

p-Cymene

p-Cymene is the biological precursor of carvacrol. p-Cymene is hydrophobic and causes swelling of the cytoplasmic membrane to a larger extent than Carvacrol [16]. p-Cymene is not an effective antibacterial when used alone [10][17], but when combined with carvacrol, synergism was observed against *B. cereus* in vitro and in rice [18]. The higher ability of p-Cymene to be incorporated in the lipid bilayer of *B.cereus* likely facilitates transport of carvacrol across the cytoplasmic membrane [16].

2.2.2 Terpinene

γ-Terpinene did not inhibit the growth of *S. typhimurium* [19], whereas α-terpinene inhibited 11 of the 25 bacterial species screened [10].

2.2.3 Terpenoids

Terpenoids are terpenes with added oxygen molecules or terpenes whose methyl groups have been moved or removed by specific enzymes [15]. Thymol, carvacrol, linalool, menthol, geraniol, linalyl acetate, citronellal and piperitone are the most common terpenoids. The antimicrobial activity of most terpenoids is related to their functional groups, and the hydroxyl group of the phenolic terpenoids and the presence of delocalised electrons are important elements for their antimicrobial action. For example, carvacrol which has an added hydroxyl group is more effective than other *p*-cymene. The exchange between the hydroxyl group and a methyl ether in carvacrol can affect its hydrophobicity and antimicrobial activity. The position of the hydroxyl group in the phenolic molecule does not affect the trend of the antimicrobial activity. Compared with carvacrol, thymol has similar antimicrobial activity against *B. cereus*, *S. aureus* and *P. aeruginosa* despite the positioning of its attached groups [16].

Examples of terpenoids:

- Carvacrol and thymol

Thymol is structurally very similar to carvacrol, but with the hydroxyl group at a different location on the phenolic ring. Both substances appear to make the cell membrane permeable [20]. Carvacrol and thymol are able to disintegrate the outer membrane of gram-negative bacteria. This brings about the release of membrane components and increases the permeability of the cytoplasmic membrane to ATP. Studies with *B. cereus* showed that carvacrol dissolved in the phospholipid bilayer of the cell membrane and is assumed to align between the fatty acid chains [18]. The distortion of the physical structure caused expansion and destabilisation of the membrane, increasing membrane fluidity, which in turn increased passive permeability [16].

The passage of *B. cereus* cell metabolites across the cell membrane on exposure to carvacrol has also been investigated. Intracellular and extracellular ATP measurements revealed that the level of ATP within the cell decreased while there was no proportional increase outside the cell. It is therefore presumed that the rate of ATP synthesis was reduced or that the rate of ATP hydrolysis was increased. Measurements of the membrane potential of microbial cells in the log phase revealed a sharp decrease when carvacrol was added and indicated a weakening of the proton motive force. The pH gradient across the cell membrane was also weakened by the presence of carvacrol and was completely dissipated in the presence of 1 mM or more. Furthermore, intracellular levels of potassium ions dropped whilst extracellular amounts increased proportionately, the total amount remaining constant [21]. It was concluded that carvacrol forms channels through the membrane by pushing apart the fatty acid chains of the phospholipids, allowing ions to leave the cytoplasm [18]. Oregano EO, containing carvacrol as a major component, causes leakage of phosphate ions from S. aureus and P. aeruginosa [20].

Apart from the growth inhibition of vegetative bacterial cells, the production of toxin is also inhibited. Carvacrol is able to inhibit the production of diarrhoeal toxin by B. cereus in broth and in soup.

Juven et al. (1994) examined the working of thymol against S. typhimurium and Staphylococcus aureus and hypothesised that thymol binds to membrane proteins hydrophobically and by means of hydrogen bonding, thereby changing the permeability characteristics of the membrane. Thymol was found to be more inhibitive at pH 5.5 than 6.5. At low pH the thymol molecule would be undissociated and therefore more hydrophobic, and so may bind better to the hydrophobic areas of proteins and dissolve better in the lipid phase [19].

- **Carvone**

Carvone (2-methyl-5-(1-methylethenyl)-2-cyclohexen-1-one) dissipated the pH gradient and membrane potential of bacterial cells. The specific growth rate of Escherichia coli, Streptococcus thermophiles and L. lactis decreased with increasing concentrations of carvone, which shows that it acts by disturbing the metabolic energy status of cells [22]. However, another study found that carvone was ineffective on the outer membrane of Escherichia coli and S. typhimurium and had no effect on the intracellular ATP pool [23]. With increasing amount of carvone, Oosterhaven et al. [22] observed a decrease in the growth rate of E. coli, Streptococcus thermophilus and L. lactis and hypothesised that the compound might act by disturbing the metabolic energy status of cells. In contrast, another study [23] found that carvone was ineffective against the OM of *E. coli* and *S. typhimurium* and did not affect their intracellular ATP pool.

2.2.4 Phenylpropenes

Phenylpropenes contain a six-carbon aromatic phenol group and a three-carbon propene tail from cinnamic acid, which is produced during the first step of phenylpropanoid biosynthesis. These compounds represent a relatively small portion of EOs. Eugenol, isoeugenol, vanillin, safrole and cinnamaldehyde are the most studied phenylpropenes [5]. Most antimicrobial activity of these molecules is conferred by their free hydroxyl groups [24]. The antimicrobial activity of the phenylpropenes also depends on the type and number of substitutions on the aromatic ring and similar to most other EOs, on the microbial strain and conditions in which the EO is tested [25].

Eugenol

Eugenol is a major component (approximately 85%) of clove oil. Sub-lethal concentrations of eugenol inhibited the production of amylase and proteases by *B. cereus*. Cell wall deterioration and a high degree of cell lysis were also noted [26]. Eugenol inhibited the production of extracellular enzyme by *Bacillus cereus*. The hydroxyl group on eugenol is thought to bind to proteins, preventing enzyme action in *E. aerogenes* [27]. The antimicrobial activity of eugenol can be ascribed to the presence of a double bond in the α, β positions of the side chain and to a methyl group located in the γ position [28]. Isoeugenol is more active against bacteria than eugenol and is also effective against yeasts and mould [24]. It is important to note that eugenol and isoeugenol interestingly exhibit higher activity against Gram-negative bacteria than Gram-positive bacteria [29]. Eugenol alters the membrane, affects the transport of ions and ATP and changes the fatty acid profile of different bacteria. It also acts against different bacterial enzymes, including ATPase, histidine carboxylase, amylase and protease [27].

Cinnamaldehyde

Cinnamaldehyde (3-phenyl-2-propenal) has been shown to be inhibitive to the growth of *E.coli* 0157:H7 and *S. typhimurium* at similar concentrations to carvacrol and thymol. It did not disintegrate the outer membrane or deplete the intracellular ATP pool [23]. The carbonyl group is thought to bind to proteins, preventing the action of amino acid decarboxylases in *E. aerogenes* [27]. Cinnamaldehyde is generally less powerful than eugenol [30], but when it is used against *E. coli* and *S. typhimurium*, its activity is similar to thymol and carvacrol, the most potent EOs [23]. At low concentrations, cinnamaldehyde inhibits enzymes involved in cytokine interactions or other less important cell functions, and at higher concentrations, it acts as an ATPase inhibitor. At a lethal concentration, it perturbs the membrane. Some studies have reported conflicting information on the membrane-perturbing activity of cinnamaldehyde. For example, a sub-lethal concentration of the molecule does not affect the integrity of the membrane in *E. coli* but can inhibit the growth and bioluminescence of the microorganism *Photobacterium leiognathi*. This suggests that cinnamaldheyde gains access to the periplasm and perhaps also to the cytoplasm [23]. Cinnamaldehyde is also capable of altering the lipid profile of the microbial cell membrane [27].

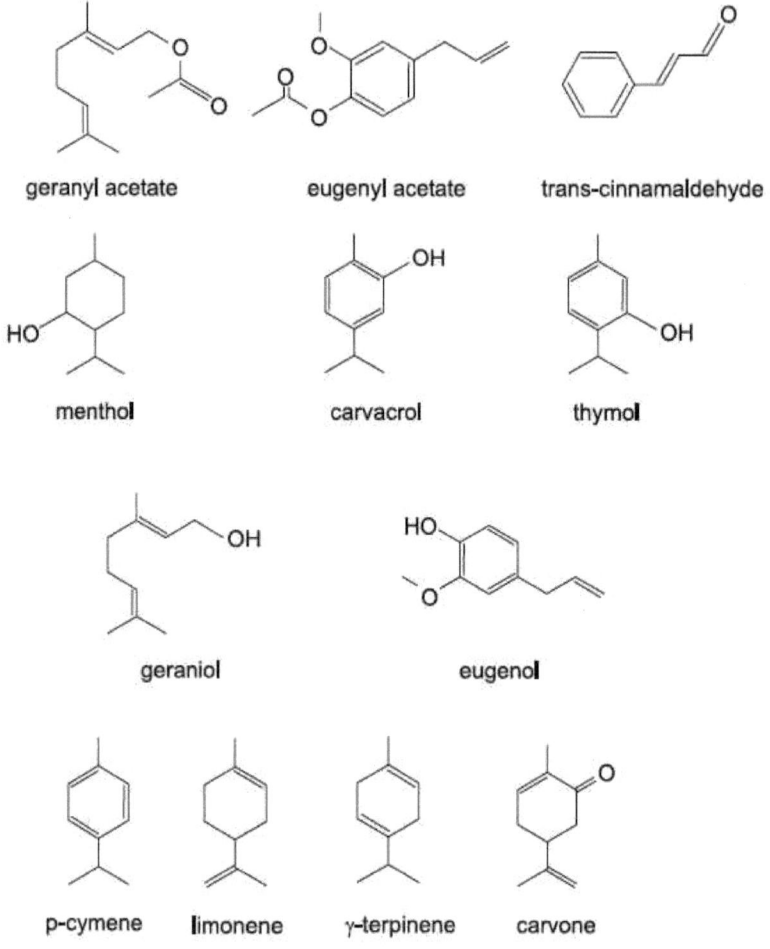

Figure 3: Structural formulae of selected components of EO [3].

2.3 Synergism and antagonism between components of Essential oils

The inherent activity of any oil is expected to be based on the chemical configuration of the components, the amount of each component present and to the interactions between them [31]. An additive effect is observed when the combined effect is equal to the sum of the individual effects. Antagonism occurs when the effect of one or both compounds is less when they are applied together than when individually applied. Synergism occurs when the effect of the combined substances is greater than the sum of the individual effects. Fractions of cilantro, coriander, dill and eucalyptus EOs (each containing several components), when mixed in various combinations, resulted in additive, synergistic or antagonistic effects [31]. A mixture of cinnamaldehyde and eugenol at 250 and 500 µg ml^{-1} respectively inhibited growth of *Staphylococcus sp., Micrococcus sp. Bacillus sp.* and *Enterobacter sp.* for more than 30 days completely, whereas the substrates applied individually did not inhibit growth [32]. Another example of the synergic effect of essential oils will be shown in subsequent chapter using the essential oils of lemon grass and orange peels.

Some studies have concluded that whole EOs have a greater antibacterial activity than mixing major components [33][34], which suggests that minor components are critical to the activity of EOs and may have a synergistic effect or potentiating influence. The two structurally similar major components of oregano EO, carvacrol and thymol, were found to give a synergistic effect when tested against *Staphylococcus aureus* and *P. aeruginosa* [20]. Synergism between carvacrol and its biological precursor p-cymene was observed when acted on the vegetative cells of *B. cereus*. It appears that p-cymene, a very weak antibacterial, swells bacterial cell membranes to a greater extent than carvacrol does. By this mechanism p-cymene probably enables carvacrol to be more easily transported into the cell so that a synergistic effect is achieved when the two are used together.

2.3.1 Synergism and antagonism between EO components and food preservatives or preservation techniques.

Some potential synergists have been suggested for use with EOs. These include; low pH, low water activity, chelators, low oxygen tension, mild heat and raised pressure. However, not all of these have been researched in foodstuffs [35]. A number of studies have demonstrated the combined effect of EOs or their components with food additives such as sodium chloride, sodium nitrite and nisin and with preservation techniques of mild heat treatment, high hydrostatic pressure and anaerobic packaging. Sodium chloride (NaCl) has been shown to work as a synergist and an antagonist under different circumstances with EOs and/or their components. Synergism between NaCl and mint oil against *S. enteritidis* and *L. monocytogenes* has been recorded in taramosalata [8]. The combined use of 2–3% NaCl and 0.5% clove powder (containing eugenol and eugenyl acetate) in mackerel muscle extract has been found to totally prevent growth and histamine production by *E. aerogenes*. The suggested mechanism for this is that eugenol increases the permeability of the cells after which NaCl inhibits growth by its action on intracellular enzymes [27]. Antagonistic effects of salt were found with carvacrol and p-cymene against *B. cereus* in rice: carvacrol and p-cymene worked synergistically, but this effect was reduced when salt was added (1.25 g l-1 rice) [18]. In the same study, soy sauce was shown to exhibit synergy with carvacrol. However, this synergy was also cancelled out by the presence of salt [18]. Salt at 4% w/v in agar did not improve the antibacterial activity of cinnamaldehyde against gram-positive and gram-negative bacteria [32].

Combinations of oregano EO with sodium nitrite have been examined for their effect on growth and toxin production by *C. botulinum*. Oregano oil acted synergistically with nitrite to inhibit growth in broth, whereas when applied singly up to 400 ppm, oregano oil had no significant inhibitive effect on growth. The simultaneous application of nisin (0.15 µg ml-1) and carvacrol or thymol (0.3 mmol l-1 or 45 µg ml-1) caused a larger reduction in viable counts for strains of *B. cereus* than was observed when the antimicrobials were individually applied. The maximum reduction of viability was achieved in cells that had experienced prior exposure to mild heat treatment at 45 $^{\circ}$C (5 min for exponentially growing cells and 40 min for stationary phase cells) [36]. At pH 7 the synergistic action of nisin and carvacrol was significantly greater at 30 $^{\circ}$C than at 8 $^{\circ}$C, which appears to indicate temperature induced changes in the permeability of the cytoplasmic membrane [36].

The combined effect of carvone (5 mmol l^{-1}) and mild heat treatment (45 $^{\circ}$C, 30 min) on exponentially growing cells of *L. monocytogenes* grown at 8 $^{\circ}$C has been studied. Separately, the two treatments demonstrated no loss in viability but a decrease of 1.3 log units in viable cell numbers was recorded when they were combined. Cells grown at 35 or 45 $^{\circ}$C were not susceptible to the same combined treatment. The authors hypothesised that the phospholipid composition of the cytoplasmic membrane of cells grown at 8 $^{\circ}$C has a higher degree of unsaturation in order to maintain fluidity and function at low temperatures. This high degree of unsaturation causes the membranes of these cells to be more fluid at 45 $^{\circ}$C than the membranes of cells grown at that temperature. This increased fluidity would enable carvone to dissolve more easily into the lipid bilayer of cells grown at 8 $^{\circ}$C than into the bilayer of cells grown at 45 $^{\circ}$C.

Chapter Three

3.1 Effects of essential oils on pathogenic microorganisms

3.1.1 Antibacterial actions of essential oils

Conner (1993) discovered that cinnamon, cloves, pimento, thyme, oregano, and rosemary plants had strong inhibitory effects against several bacterial pathogens. Kim *et al* (1995) [37] reported that essential oils extracted from some medicinal plants had antibacterial effects against all the five food borne pathogens tested. This result was attributed to the presence of phenolic compounds such as carvacrol, eugenol and thymol [38]. Ouattara *et al.*, (1997) [9] observed the antibacterial activity of selected spices on the meat spoilage bacteria. More so, Arora and Kaur (1999) [38] analyzed the antimicrobial activity of garlic, ginger, clove, black pepper and green chilli on the human pathogenic bacteria viz. *Bacillus sphaericus, Enterobacter aerogenes, E. coli, P. aeruginosa, S. aureus, S. epidermidis, S. typhi* and *Shiguella flexneri* and stated that amongst all the tested spices, aqueous garlic extracts was sensitive against all the bacterial pathogens. Similarly, effect of clove extracts on the production of verotoxin by enterohemorrhagic *Escherichia coli* O157:H7 was investigated by Sakagami *et al.*, (2000) [39] and it was shown from the study that the verotoxin production was inhibited by clove extract. However, Elgayyar *et al.*, (2001) [40] examined the effectiveness of cardamom, anise, basil, coriander, rosemary, parsley, dill and angelica essential oil for controlling the growth and survival of pathogenic and saprophytic microorganisms. The results of their study showed that only essential oils extracted from oregano, basil and coriander plants had inhibitory effect against *Pseudomonas aeruginosa, S. aureus* and *Yersinia enterocolitica*. The effect of oregano essential oils on the behaviour of *Salmonella typhimurium* in sterile and naturally contaminated beef fillets stored under aerobic and modified atmospheres was also observed [41]. They concluded that the addition of oregano essential oils reduced the initial population of the tested bacterial pathogens. Hood *et al.*, (2003) [42] reported that bacterial growth may be inhibited by the ample application of essential oils or their use at high concentrations and their mode of action results in decline of the growth of bacterial cells. Similarly, Sokovic *et al.*, (2009) [43] observed the antibacterial activity of essential oils extracted from thyme and mint leaves against the *Staphylococcus aureus, Salmomella typhimurium* and *Vibrio parahaemolyticus*. The result showed that all the plants have antibacterial activity against the tested pathogens but the effect of thyme leaves extract was greater than the other plant. Table 2 shows a few plants from which essential oils could be extracted, the parts involved and the microorganisms inhibited.

Table 2. Plants containing essential oils and inhibited microorganisms.

Plant	Part used	Chemical compounds	Inhibited microorganisms
Cymbopogon citrates *Allium sativum*	Friuts Bulb	Ethanolic compounds Isothiocynate	*Enteriobacteriacae* *Candida spp*
Thymus vulgaris	Ariel part	Thymol, linalool, carvacrol	*L.monocytogenes, E.coli, S.aureus*
Pimpinella anisum	Seed	Trans-anethole	*S.typhimirium, E.coli*
Origanum vulgare	Ariel part	Carvacrol, Thymol, γ-Terpinene	*L.monocytogenes, E.coli,* Adeno virus, Polio virus
Feoniculum vulgare	Seed	Trans-anethole	*Fusarium oxysporium, Aspergillus flavus*
Rosmarinus officinalis	Flower	Benzaylacetate, linalool, α-pinene	*E.coli, S.typhimurium, B.cereus, S.aureus*
Thymus vulgaris, Mentha piperita	Ariel part	1,8-Cineole, Eugenel	*S.aureus, E.coli*

3.1.2 Antifungal actions of essential oils

Essential oils and their components have been used broadly against moulds. The essentials oils extracts from many plants such as basil, citrus, fennel, lemon grass, oregano, rosemary and thyme have shown their considerable antifungal activity against the wide range of fungal pathogens [44]. Arora and Kaur (1999) [38] observed the sensitivity essential oil of spices against some fungal pathogens and concluded that garlic and clove extracts had strong ability to inhibit the growth of *Candida acutus, C. albicans, C. apicola, C. catenulata. C. inconspicua, C. tropicalis, Rhodotorula rubra, Sacharomyces cerevisae* and *Trignopsis variabilis*. However, Delaquis and Mazza (1995) reported the antimicrobial effects of isothiocyanate isolated from onion and garlic plants and stated that isothiocyanates may inactivate the extracellular enzymes through the oxidative cleavage of disulphide bonds. According to the report of Ultee and Smid (2001) [7], oregano and thyme essential oils are apparently amongst the best inhibitors of fungal pathogens because of the presence of the phenolic compounds such as carvacrol and thymol as main constituents which might be involved in the disruption of fungal cell membrane. Antifungal activity of essential oil and its derivatives has been studied on viable cells count, mycelia growth and mycotoxins producing ability of moulds and amongst all tested essential oils, clove, cinnamon and oregano essential oils were effective against *Aspergillus parasiticus* and *Fusarium moniliforme* [45].

3.1.3 Antiviral actions of essential oils

The antiviral activity of essential oils were tested against many enveloped RNA and DNA viruses, such as herpes simplex virus type 1 and type 2 (DNA viruses), dengue virus type 2 (RNA virus), and influenza virus (RNA virus). However, essential oils extracted from oregano and clove were also tested against non-enveloped RNA and DNA viruses, such as adenovirus type 3 (DNA virus), poliovirus (RNA virus), and coxsackievirus B1 (RNA virus) [46]. *Melissa officinalis* essential oils could also inhibit the replication of HSV-2, due to the presence of citral and citronellal [47]. The ability of replication of HSV-1 could be suppressed by the *in-vitro* incubation of various essential oils. It is obvious from the study of Sinico *et al.*, (2005) [48] that the Herpes simplex virus type 1 (HSV-1), which is the cause of common viral infections in humans, such as mucocutaneous herpes infections, herpetic keratitis, herpetic encephalitis, and neonatal herpes could be strongly suppressed by the activity of essential oils extracted from *Artemisia arborescens*. Thus, it has been concluded that the essential oils have been frequently used as antiviral agents against several viral diseases in human [49] and it has a potential to be used as alternative to synthetic antiviral drugs [50][51].

3.1.4 A summary of the mechanism of action of EOs

Hydrophobicity of EOs and their components is an important characteristic that enables them to partition in the lipids of the bacterial cell membrane and mitochondria, disturbing the structures and rendering them more permeable [52]. Leakage of ions and other cell contents can then occur [53][16]. Although a certain amount of leakage from bacterial cells may be tolerated without loss of viability, extensive loss of cell contents or the exit of critical molecules and ions will lead to cell death [54]. Evidence from studies with tea tree oil and *E. coli* have shown that cell death may occur before lysis [55].

Generally, the EOs that possess the strongest antibacterial properties against food borne pathogens contain a high percentage of phenolic compounds such as carvacrol, eugenol (2-methoxy-4-(2-propenyl) phenol) and thymol [10][17][20]. It is therefore reasonable to conclude that their mechanism of action would be similar to other phenolics. This is generally considered to be the disturbance of the cytoplasmic membrane, disrupting the proton motive force (PMF), electron flow, active transport and coagulation of cell contents [56][57].

As far as non-phenolic components of EOs are concerned, the type of alkyl group has been found to influence activity. For example, limonene (1-methyl-4-(1-methylethenyl) cyclohexene) is more active than p-cymene [10].

Components of EO also appear to act on cell proteins embedded in the cytoplasmic membrane [58]. Enzymes such as ATPases are known to be located in the cytoplasmic membrane and to be bordered by lipid molecules. Two possible mechanisms have been suggested whereby cyclic hydrocarbons could act on proteins.

- Lipophilic hydrocarbon molecules could accumulate in the lipid bilayer and distort the lipid–protein interaction; alternatively,

- Direct interaction of the lipophilic compounds with hydrophobic parts of the protein is also possible [19][56].

Chapter Four

4.1 Tests of Antimicrobial activity of EOs in food systems

Despite that EOs perform very well in antibacterial assays in vitro, it has generally been found that a greater concentration of EO is needed to obtain the same effect in foods [59].

The ratio has been recorded to be approximately two fold in semi-skimmed milk [60], 50-fold in soup [7] 10-fold in pork liver sausage [61], and 25- to 100-fold in soft cheese [62]. An exception to this phenomenon is *Aeromonas hydrophila*; in which no greater proportion of EO was needed to inhibit this species on cooked pork and on lettuce in comparison to tests in vitro [63][64]. The greater availability of nutrients in foods compared to laboratory media may enable bacteria to repair damaged cells faster [33] which could limit the result in food. Not only are the intrinsic properties of the food (fat/protein/water content, antioxidants, preservatives, pH, salt and other additives) relevant in this respect—the extrinsic determinants (temperature, packaging in vacuum/gas/air, characteristics of microorganisms) can also influence bacterial sensitivity [8].

Generally, the susceptibility of bacteria to the antimicrobial effect of EOs appears to increase with a decrease in the pH of the food, the storage temperature and the amount of oxygen within the packaging [8][65][66]. At low pH the hydrophobicity of an EO increases, enabling it to more easily dissolve in the lipids of the cell membrane of target bacteria [19].

It is generally supposed that the high levels of fat and/or protein in foodstuffs protect the bacteria from the action of the EO in some way [61][8]. It is also possible that lower water content of food compared to laboratory media may hamper the progress of antibacterial agents to the target site in the bacterial cell [67]. Mint oil in the high fat products exhibited little antibacterial effect against *L. monocytogenes* and *S. enteritidis*, whereas in cucumber and yoghurt salad (low fat) the same EO was much more effective [8]. Although the improved effectiveness in cucumber and yoghurt salad may be partly attributed to the low pH (4.3). This would seem to indicate that fat percentage might exert a greater influence on the antibacterial effect of EOs as well as the pH value.

Protein content has also been put forward as a factor inhibiting the action of clove oil on *Salmonella enteritidis* in diluted low-fat cheese [67]. Carbohydrates in foods do not appear to protect bacteria from the action of EOs as much as fat and protein do [68]. A high water and/or salt level facilitates the action of EOs [8][65]. The physical structure of a food may limit the antibacterial activity of EO. A study of the relative performance of oregano oil against *S. typhimurium* in broth and in gelatine gel revealed that the gel matrix dramatically reduced the inhibitory effect of the oil. This was presumed to be due to the limitation of diffusion by the structure of the gel matrix [65].

4.2 Effects of EOs against pathogens in food samples

4.2.1 Meat and meat products

Some EOs stand out as better antibacterial agents in certain foods than others. Certain oils act as better antibacterial than others for meat applications. A number of studies showed that eugenol and coriander, clove, oregano and thyme oils were effective at levels of 5–20 µl g^{-1} in inhibiting *L. monocytogenes, A. hydrophila* and autochthonous spoilage flora in meat products, sometimes causing a marked initial reduction in the number of recoverable cells [66][69]. Mustard, cilantro, mint and sage oils were less effective or ineffective [8][33][70]. The high fat content of meat appears to significantly reduce the action of EOs in meat products [8][33]. For example, mint and cilantro EOs were not effective in products with a high level of fat, such as pate′ (which generally contains 30–45% fat) and a coating for ham containing canola oil Immobilising cilantro EO in a gelatine gel, however, improved the antibacterial activity against *L. monocytogenes* in ham [33]. A study found that encapsulated rosemary oil was much more effective than standard rosemary EO against *L.*

monocytogenes in pork liver sausage, although whether the effect was due to the encapsulation or the greater percentage level used was not further elucidated [61]. The activity of oregano EO against Clostridium botulinum spores has been studied in a vacuum packed and pasteurised minced (ground) pork product. Concentrations of up to 0.4 µl g^{-1} oregano EO were found not to significantly influence the number of spores or to delay growth. However, in the presence of low levels of sodium nitrite which delayed growth of bacteria and swelling of cans when applied alone, the same concentration of oregano EO enhanced the delay. The delay of growth was dependent on the number of inoculated spores; at 300 spores g^{-1} the reduction was greater than at 3000 spores g^{-1} [71].

4.2.2 Fish dishes

As stated earlier, high fat content appears to reduce the effectiveness of antibacterial EOs in fatty food items such as fish. For example, oregano oil at 0.5 µl g-1 is more effective against the spoilage organism. Oregano oil is more effective on fish than mint oil, even in fatty fish dishes; this was confirmed in two experiments with fish roe salad using the two EOs at the same concentration (5–20 µl g^{-1}) [8][72]. The spreading of EO on the surface of whole fish or using EO in a coating for shrimps appears effective in inhibiting the respective natural spoilage flora [9][73].

4.2.3 Dairy products

Mint oil at 5–20 µl g^{-1} is effective against *S. enteritidis* in low fat yoghurt and cucumber salad [8]. Mint oil has been shown to inhibit the growth of yoghurt starter culture species at 0.05–5 µl g^{-1} but cinnamon, cardamom and clove oils have been reported to be much more effective [74].

4.2.4 Vegetables

Vegetables generally have a low fat content. In vegetable dishes just as for meat products, the antimicrobial activity of EOs is benefited by a decrease in storage temperature and/or a decrease in the pH of the food [65]. This may contribute to the positive results obtained with EOs. All EOs and their components that have been tested on vegetables appear effective against the natural spoilage flora and food borne pathogens at levels of 0.1–10 µl g^{-1} in washing water [75]. Cinnamaldehyde and thymol are effective against six Salmonella serotypes on alfalfa seeds when applied in hot air at 50 °C as fumigant. Increasing the temperature to 70 °C reduced the effectiveness of the treatments. This may be due to the volatility of the compounds. Oregano oil effectively inhibited *Escherichia coli* O157:H7 at 7–21 µl g^{-1} and reduced final populations in eggplant salad compared to the untreated control.

4.2.5 Fruits

Carvacrol and cinnamaldehyde were very effective at reducing the viable count of the natural flora on kiwifruit when used at 0.15 µl ml^{-1} in dipping solution, but less effective on honeydew melon. It is possible that this difference has to do with the difference in pH between the fruits; the pH of kiwifruit was 3.2–3.6 and that of melon was 5.4–5.5 [76]. The lower the pH, the more effective EOs and their components generally are.

Chapter Five

5.1 An experimental approach using the essential oils of Lemon grass and Orange peels

5.1.1 Extraction of essential oil

Essential oils are often volatile oils. It is therefore important to ensure that fresh plant materials are used for oil extraction and not dried samples. For this investigation to determine the effect of the essential oils of lemon grass and orange peels against some microbes, freshly harvested lemon grass were used as well as freshly peeled oranges. In case of interruptions after harvesting or before the extraction process, the leaves and peels should be stored in a refrigerator. The extraction of EOs can be carried out through hydro or steam distillation. It is however important to note that the amount of essential oils present in plants vary from one plant to another. For instance, the amount of peels required to obtain 20 ml of orange peel essential oil is not as much as that required for the same amount of lemon grass oil. A larger quantity, in fact five- fold more is required to obtain the same quantity for lemon grass oil. Oil yield can also be affected by seasonal changes aside quantity of plant parts. After oil extraction, it is important to store essential oils in amber bottles to avoid degradation by exposure to light. Refrigeration after bottling is a good way of storing EOs.

5.2 Preparation of graded concentration of the samples by serial dilution.

- Pipette 2 ml of the essential oils of lemon grass and orange peels into different test tubes labelled 1 in each case (100%).
- Pipette 1 ml of Tween 20 (solvent for dissolving essential oils) into five test tubes labelled 2 to 6. This should be done in two categories since two essential oils are involved.
- Pipette 1 ml of pure essential oil into tube 2 from tube 1 (50%).
- Pipette 1 ml from tube 2 into tube 3 (25%).

- Pipette 1 ml from tube 3 into tube 4 (12.5%).
- Pipette 1 ml from tube 4 into tube 5 (6.25%).
- Pipette 1 ml form tube 5 into tube 6 (3.12%).
- Test tubes 7 and 8 are for negative and positive control respectively. For instance, test tube 7 should contain 1 ml Tween 20 (solvent used) while test tube 8 should contain an antibiotic, for instance gentamycine, ampicillin etc. for bacteria, ticonazole for experiment involving fungi.

For this study, bacteria pathogens were used. These are *Staphylococcus epidermidis, Staphylococcus aureus, Lactobacillus casei, Lactobacillus brevis.* The Staphylococcus species used are pathogenic while the lactobacillus are not pathogenic. Ingestion of *L. brevis* has been shown to improve human immune function, and has been pated several times as well as *L. casei.* The microbes used were obtained from food sources.

5.2.1 Pour Plate Method (Bacteria)

- Prepare an overnight culture of each organism by taking a loop full of the organism from stock.
- Inoculate each into the sterile nutrient broth of 5 ml each.
- Incubate for 18-24 hours at 37^0C.
- Obtain 0.1 ml of each organism from the overnight culture and put into 9.9 ml of sterile distilled water to get 1: 100 of the dilution of the organisms.
- Take 0.2 ml from the diluted organisms into prepared sterile nutrient agar at 45^0C.
- Aseptically pour into sterile petri dishes.
- Allow to solidify for about 45-60 minutes.
- Make wells using a sterile cork borer of 8 mm according to the number of graded concentration of the samples.

- In each well, produce the graded concentrations of the samples in duplicate.
- Allow the plates to stay on the bench for about 2 hours to allow pre-diffusion.
- Incubate the plates uprightly in the incubator for 18-24 hours at 37^0C.
- Observe the bacteria plates after 24 hours of incubation.
- Clear zones of inhibition would be observed on some plates with higher concentration of essential oil while shorter or no zone at all for lower concentrations. The zone of inhibition is measured in mm.

5.2.2 Minimum Inhibitory Concentration (MIC) and Minimum Bactericidal Concentration (MBC)

The MIC for each EO was determined using the first four concentrations. Each of the plates used was divided into four parts representing the four microorganisms after solidification of the nutrient agar. Since four concentrations of the initial six were considered for MIC, a total of four plates was used in this case. The growth of microbes in each of the plates was checked for after 24 hours. Divisions that showed no growth were then transferred into another set of plates for Minimum Bactericidal Concentration (MBC). The MBC is to determine if the microbe that did not grow at the end of the MIC were dead or alive. The MIC give the minimum inhibition concentration which means that the growth of the organisms have been inhibited but not necessarily dead. MBC gives the minimum concentration at which the antibacterial agent has completely killed the microorganism or pathogen. After 24 hours of MBC, organisms that show no growth were considered dead confirming the bactericidal property of the antibacterial used or essential oil.

5.3 Results

Table 3

Table showing Inhibition zone using the essential oil of orange peels to inhibit the growth of the four microorgansims used.

Concentration	S.epidermidis (mm)	S. aureus (mm)	L. casei (mm)	L. brevi (mm)
100%	14	20	14	14
50%	12	18	12	12
25%	10	16	10	10
12.5%	-	14	-	-
6.25%	-	12	-	-
3.12%	-	10	-	-

Table 4

Table showing the inhibition zone using the essential oil of lemon grass to inhibit the growth of the four microorganisms used.

Concentration	S.epidermidis (mm)	S. aureus (mm)	L. casei (mm)	L. brevis (mm)
100%	20	24	14	14
50%	18	20	12	12
25%	14	18	10	10
12.5%	12	16	-	-
6.25%	10	14	-	-
3.12%	-	10	-	-
Negative control	10	12	10	10
Positive control	38	40	36	38

'-'means no zone of inhibition.

Table 5

Table showing the Minimum Inhibitory Concentration (MIC) of orange peels essential oil.

Concentration	S. epidermidis	S. aureus	L. casei	L. brevis
100%	-	-	-	-
50%	+	-	+	+
25%	+	-	+	+
12.5%	+	+	+	+

Table 6

Table showing the Minimum Inhibition Concentration (MIC) of lemon grass essential oil.

Concentration	S. epidermidis	S. aureus	L. casei	L. brevis
100%	-	-	+_	+_
50%	-	-	+	+
25%	-	-	+	+
12.5%	+	-	+	+

'-' means no growth while '+' means growth observed and '+_' means unclear growth.

Table 7

Table showing the Minimum Bactericidal Concentration (MBC).

Concentration	S. epidermidis	S. aureus	L. casei	L. brevis	Essential oil
100%	+	-	+	+	Orange
50%	+	+	+	+	Orange
100%	-	-	+	+	Lemon
50%	-	-	+	+	Lemon
25%	+	+	+	+	Lemon

'-' means no growth while '+' means growth observed.

5.4 Discussion

From the results obtained, the essential oil of lemon grass showed a higher level of microbial growth inhibition and bactericidal property compared to orange peel essential oil. The antibacterial property of the oils increased as concentration increased. According to tables 3 and 4, subtracting the value of the negative control from the oil value showed that even at the highest concentration, the essential oil of orange peel had a lower antibacterial effect based on the short inhibition zone compared to lemon grass essential oil. According to tables 5 and 6, the MIC varies from one organism to another. Both oils showed no growth inhibition for *L. casei and L. brevis* except for the 100% orange essential oil. The inhibition of the other organisms was led by the essential oil of lemon grass especially from 100% to 25%

47

concentration. According to table 7, the MBC for orange peel essential oil is 100% for S. aureus only while that of lemon grass is 50%. Hence the essential oil of lemon grass can effectively mitigate the growth (bacteriostatic) of *S. epidermidis* and *S. aureus* at 25% concentration or more (MIC) and cause death (bactericidal) at a minimum concentration (MBC) of 50%. Lemon grass essential oil is therefore a potent antibacterial agent with good potential for use in food pathogen mitigation.

Conclusion

The structural composition and the functional groups present in essential oils play important role in determining the antimicrobial activity. Essential oils contain a variety of volatile molecules such as terpenes and terpenoids, phenol-derived aromatic and aliphatic compounds, which have bactericidal, virucidal, and fungicidal consequences. Essential oils directly affect the cell membrane of pathogenic microorganisms by causing an increase in permeability and leakage of vital intracellular constituents. Eventually resulting into microbial cell death. Therefore, essential oils extracts from the plants reviewed possess the potential for use as alternative natural antimicrobial substances. Hence, essential oils have the potential for use in food pathogen mitigation. As demonstrated by the essential oil of lemon grass, the effectiveness of the oil against *S. epidermidis* and *S. aureus* can be used to mitigate the growth of these bacterial agents that cause spoilage or deterioration of food.

References

1. World Health Organization (2003). *The present state of foodborne disease in OECD countries*. Geneva, Switzerland: World Health Organization.

2. Philippe, S., Souaïbou, F., and Dominique, S. (2012). Major component and potential applications of plant essentials oils as natural food preservatives: a short review research results, *2*, 45–57.

3. Burt, S. (2004). Essential oils: their antibacterial properties and potential applications in foods—a review, *94*, 223– 253.

4. *The World Health Report: reducing risks, promoting healthy life.* (2002). Geneva: World Health Organization (WHO).

5. Filomena, N., Florinda, F., Laura, D. M., Raffaele, C., & Vincenzo, D. F. (2013). Effect of Essential Oils on Pathogenic Bacteria. Pharmaceuticals (*6*) 1451-1474; doi:10.3390/ph6121451.

6. Mari, M., Bertolini, P., and Pratella, G. C. (2003). Non-conventional methods for the control of post-harvest pear diseases. *Journal of Applied Microbiology*, *94*(5), 761–766.

7. Ultee, A., and Smid, E. J. (2001). Influence of carvacrol on growth and toxin production by Bacillus cereus. *International Journal of Food Microbiology*, *64*(3), 373–378.

8. Tassou, C. C., Drosinos, E. H., and Nychas, G. J. (1995). Effects of essential oil from mint (Mentha piperita) on Salmonella enteritidis and Listeria monocytogenes in model food systems at 4 degrees and 10 degrees C. *The Journal of Applied Bacteriology*, *78*(6), 593–600.

9. Ouattara, B., Simard, R. E., Holley, R. A., Piette, G. J., & Bégin, A. (1997). Antibacterial activity of selected fatty acids and essential oils against six meat spoilage organisms. *International Journal of Food Microbiology*, *37*(2-3), 155–162.

10. Dorman, H. J. and Deans, S. (2000). Antimicrobial agents from plants: antibacterial activity of plant volatile oils, *88*, 08–316.

11. Bakkali F, Averbeck S, Averbeck D, Idaomar M (2008). Biological effects of essential oils- A review. Food Chem. Toxicol. 46(2): 446-475.

12. Cowan, M. M. (1999). Plant products as antimicrobial agents. *Clinical Microbiology Reviews, 12*(4), 564–582.

13. Akthar, M. S., Birhanu, D., and Tanweer, A. (2014). Antimicrobial activity of essential oils extracted from medicinal plants against the pathogenic microorganisms: A review, *2*, 001–007.

14. Cosentino, S., Tuberoso, C. I. G., Pisano, B., Satta, M., Mascia, V., Arzedi, E., and Palmas, F. (1999). In-vitro antimicrobial activity and chemical composition of Sardinian Thymus essential oils. *Letters in Applied Microbiology, 29*(2), 130–135. http://doi.org/10.1046/j.1472-765X.1999.00605.x

15. Caballero, B., Trugo, L., and Finglas, P. (2003). *Encyclopedia of food sciences and nutrition 10. 10.* Amsterdam [u.a.]: Academic Press.

16. Ultee, A., Bennik, M. H. J., & Moezelaar, R. (2002). The Phenolic Hydroxyl Group of Carvacrol Is Essential for Action against the Food-Borne Pathogen Bacillus cereus. *Applied and Environmental Microbiology, 68*(4), 1561–1568. http://doi.org/10.1128/AEM.68.4.1561-1568.2002

17. Juliano, C., Mattana, A., Usai, M., (2000). Composition and in vitro antimicrobial activity of the essential oil of Thymus herba-barona Loisel growing wild in Sardinia. Journal of Essential Oil Research 12, 516–522.

18. Ultee, A., Kets, E. P., Alberda, M., Hoekstra, F. A., and Smid, E. J. (2000). Adaptation of the food-borne pathogen Bacillus cereus to carvacrol. *Archives of Microbiology, 174*(4), 233–238.

19. Juven, B. J., Kanner, J., Schved, F., & Weisslowicz, H. (1994). Factors that interact with the antibacterial action of thyme essential oil and its active constituents. *The Journal of Applied Bacteriology, 76*(6), 626–631.

20. Lambert, R. J., Skandamis, P. N., Coote, P. J., and Nychas, G. J. (2001). A study of the minimum inhibitory concentration and mode of action of oregano

essential oil, thymol and carvacrol. *Journal of Applied Microbiology*, *91*(3), 453–462.

21. Ultee, A., Slump, R.A., Steging, G., Smid, E.J. (1999) Antimicrobial activity of carvacrol toward Bacillus cereus on rice. Journal of Food Protection 63 (5), 620–624.

22. Oosterhaven, K., Poolman, B., Smid, E.J. (1995). S-carvone as a natural potato sprout inhibiting, fungistatic and bacteriostatic compound. Industrial Crops and Products 4, 23–31.

23. Helander, I. M., Alakomi, H.-L., Latva-Kala, K., Mattila-Sandholm, T., Pol, I., Smid, E. J., Von Wright, A. (1998). Characterization of the Action of Selected Essential Oil Components on Gram-Negative Bacteria. *Journal of Agricultural and Food Chemistry*, *46*(9), 3590–3595. http://doi.org/10.1021/jf980154m.

24. Laekeman, G. M., van Hoof, L., Haemers, A., Berghe, D. A. V., Herman, A. G., and Vlietinck A. J. (1990). Eugenol a valuable compound forin vitro experimental research and worthwhile for further in vivo investigation. *Phytotherapy Research*, *4*(3), 90–96. http://doi.org/10.1002/ptr.2650040304.

25. Pauli, A. and Kubeczka, K.H. (2010). Antimicrobial properties of volatile phenylpropanes. *Nat. Prod. Commun.*, *5*, 1387–1394.

26. Thoroski, J., Blank, G., Biliaderis, C. (1989). Eugenol induced inhibition of extracellular enzyme production by Bacillus cereus. Journal of Food Protection 52 (6), 399– 403.

27. Wendakoon, C.N. and Sakaguchi, M. (1995). Inhibition of amino acid decarboxylase activity of *Enterobacter aerogenes* by active components in spices. *J. Food Prot. 58*, 280–283.

28. Jung, H.G. and Fahey, G.C. (1983). Nutritional implications of phenolic monomers and lignin: A review. *J. Anim. Sci.*, *57*, 206–219.

29. Hyldgaard, M., Mygind, T., and Meyer, R. L. (2012). Essential Oils in Food Preservation: Mode of Action, Synergies, and Interactions with Food Matrix

Components. *Frontiers in Microbiology, 3.* http://doi.org/10.3389/fmicb.2012.00012.

30. Gill, A. O. and Holley, R. A. (2004). Mechanisms of Bactericidal Action of Cinnamaldehyde against Listeria monocytogenes and of Eugenol against L. monocytogenes and Lactobacillus sakei. *Applied and Environmental Microbiology, 70*(10), 5750–5755. http://doi.org/10.1128/AEM.70.10.5750-5755.2004.

31. Delaquis, P.J., Stanich, K., Girard, B., Mazza, G. (2002). Antimicrobial activity of individual and mixed fractions of dill, cilantro, coriander and eucalyptus essential oils. International Journal of Food Microbiology 74, 101–109.

32. Moleyar, V. and Narasimham, P. (1992). Antibacterial activity of essential oil components. International Journal of Food Microbiology 16, 337–342.

33. Gill, A.O., Delaquis, P., Russo, P., Holley and R.A., 2002. Evaluation of anti-listerial action of cilantro oil on vacuum packed ham. International Journal of Food Microbiology 73, 83–92.

34. Mourey, A., Canillac, N. (2002). Anti-Listeria monocytogenes activity of essential oils components of conifers. Food Control 13, 289–292.

35. Gould, G.W. (1996). Industry perspectives on the use of natural antimicrobials and inhibitors for food applications. Journal of Food Protection, 82–86.

36. Periago, P.M. and Moezelaar, R. (2001). Combined effect of nisin and carvacrol at different pH and temperature levels on the variability of different strains of Bacillus cereus. International Journal of Food Microbiology 68, 141–148.

37. Kim J, Marshall MR, Wei C (1995). Antibacterial activity of some essential oils components against five foodborne pathogens. J. Agric. Food Chem. 43(11): 2839-2845.

38. Arora D.S. and Kaur J. (1999). Antimicrobial activity of spices. International Journal of Antimicrob Agents 12:257-262 20.

39. Sakagami Y, Kaioh S, Kajimura K, Yokoyamma H (2000). Inhibitory effect of clove extract on vero-toxin production by enterohemorrhagic *Escherichia coli* 0157:H7. Biocontr. Sci. 5(1): 47-49.

40. Elgayyar M, Draughom FA, Golden DA, Mount JR (2001). Antimicrobial activity off essential oils from plants against selected pathogenic and saprophytic microorganisms. J. Food Prot. 64(7): 1019-1024.

41. Sakandamis P, Tsigarida E and Nichas G.J.E (2002). The effect of oregano essential oil on survival/death of *Salmonella typhimurium* in meat stored at 5°C under aerobic, VP/MAP conditions. Food Microbiol. 19(1): 97-103.

42. Hood JR, Wilkinson JM, Cavanagh HMA (2003). Evaluation of common antibacterial screening methods utilized in essential oil research. J. Essen. Oil Res. 15(6): 428-433.

43. Sokovic MD, Vukojevic J, Marin PD, Brkic DD, Vajs V, van Griensven LJ (2009). Chemical composition of essential oils of *Thymus* and *Mentha* species and their antifungal activities. Molecules 14(1): 238-249.

44. Kivanc M, Akgul A. and Dogan A (1991). Inhibitory and stimulatory effects of cumin, oregano and their essential oils on growth and acid production of *Lactobacillus plantarum* and *Leuconostoc mesenteroides*. Int. J. Food Microbiol. 13(1): 81-85.

45. Juglal S, Govinden R and Odhav B (2002). Spices oils for the control of co-occurring mycotoxin producing fungi. J. Food Protect. 65(4): 638-687.

46. Wagstaff A, Faulds D, Gona KL (1994). Acyclovir. A reappraisal of its antiviral activity, pharmocokinetic properties and therapeutic efficacy. Drugs 47(1): 153-205.

47. Allahverdiyev A, Duran N, Ozguven M, Koltas S (2004). Antiviral activity of the volatile oils of *Melissa officinalis* L., against Herpes simplex virus type-2. Phytomedicine 11(7-8): 657-661.

48. Sinico C, De Logu A, Lai F, Valenti D, Manconi M, Loy G, Bonsignore L, Fadda AM (2005). Liposomal incorporation of *Artemisia arborescens* L.

essential oil and *in- vitro* antiviral activity. Eur. J. Pharm. Biopharm. 59(1): 161-168.

49. Koch C, Reichling J, Schnitzler P (2008). Essential oils inhibit the replication of herpes simplex virus type 1 (HSV-1) and type 2 (HSV-2). In: Preedy VR, Watson RR, (Eds.). Botanical Medicine in Clinical Practices (pp. 192-197). USA: Wallingsford.

50. Baqui AM, Kelley JI, Jabra-Rizk MA, DePaola LG, Falkler WA, Meiller, TF (2001). *In-vitro* effects of oral antiseptics human immune deficiency virus-1 and herpes simplex virus type 1. J. Clin. Periodontol. 28(7): 610-616.

51. Primo V, Rovera M, Zanon S, Oliva M, Demo M, Daghero J, Sabini L (2001). Determination of the antibacterial and antiviral activity of the essential oil from *Minthostachys verticillata* (Griseb.) Epling. Rev. Argent. Microbiol. 33(2): 113-117.

52. Sikkema, J.; Weber, F.J.; Heipieper, H.J. and de Bont, J.A. (1994) Cellular toxicity of lipophilic compounds: Mechanisms, implications, and adaptations. *Biocatalysis, 10*, 113–122.

53. Carson CF, Mee BJ, Riley TV (2002). Mechanism of action of *Melaleuca alternifolia* (Tea tree) on *Staphylococcus aureus* determined by time-kill, leakage and salt tolerance assays and electron microscopy. Antimicrob. Agents Chemother. 46(6): 1914-1920.

54. Denyer, S.P. and Hugo, W.B. (1991). Biocide-induced damage to the bacterial cytoplasmic membrane. In: Denyer, S.P., Hugo, W.B. (Eds.), Mechanisms of Action of Chemical Biocides. The Society for Applied Bacteriology, Technical Series No 27. Oxford Blackwell Scientific Publication, Oxford, pp. 171– 188.

55. Gustafson, J.E., Liew, Y.C., Chew, S., Markham, J.L., Bell, H.C., Wyllie, S.G., Warmington, J.R. (1998). Effects of tea tree oil on Escherichia coli. Letters in Applied Microbiology 26, 194–198.

56. Sikkema, J., De Bont, J.A.M. and Poolman, B. (1995). Mechanisms of membrane toxicity of hydrocarbons. Microbiological Reviews 59 (2), 201–222.

57. Davidson, P.M. (1997). Chemical preservatives and natural antimicrobial compounds. In: Doyle, M.P., Beuchat, L.R., Montville, T.J. (Eds.), Food Microbiology: Fundamentals and Frontiers. ASM, Washington, pp. 520–556.

58. Knobloch K., P. Pauli, B. Iberi, H. Weigard and N. Weis (1989). Antibacterial and antifungal properties of essential oil components. Journal of Essential oil Res., 1: 603-608.

59. Smid, E.J., Gorris, L.G.M. (1999). Natural antimicrobials for food preservation. In: Rahman, M.S. (Ed.), Handbook of Food Preservation. Marcel Dekker, New York, pp. 285–308.

60. Karatzas, A.K., Kets, E.P.W., Smid, E.J., Bennik, M.H.J. (2001). The combined action of carvacrol and high hydrostatic pressure on Listeria monocytogenes Scott A. Journal of Applied Microbiology 90, 463– 469.

61. Pandit, V.A., Shelef, L.A. (1994). Sensitivity of Listeria monocytogenes to rosemary (Rosmarinus officinalis L.). Food Microbiology 11, 57– 63.

62. Mendoza-Yepes, M.J., Sanchez-Hidalgo, L.E., Maertens, G., Marin-Iniesta, F. (1997). Inhibition of Listeria monocytogenes and other bacteria by a plant essential oil (DMC) Spanish soft cheese. Journal of Food Safety 17, 47– 55.

63. Stecchini, M.L., Sarais, I., Giavedoni, P. (1993). Effect of essential oils on Aeromonas hydrophila in a culture medium and in cooked pork. Journal of Food Protection 56 (5), 406– 409.

64. Wan, J., Wilcock, A., Coventry, M.J. (1998). The effect of essential oils of basil on the growth of Aeromonas hydrophila and Pseudomonas fluorescens. Journal of Applied Microbiology 84, 152–158.

65. Skandamis, P.N., Nychas, G.-J.E. (2000). Development and evaluation of a model predicting the survival of Escherichia coli O157:H7 NCTC 12900 in homemade eggplant salad at various temperatures, pHs and oregano essential

oil concentrations. Applied and Environmental Microbiology 66 (4), 1646–1653.

66. Tsigarida, E., Skandamis, P., Nychas, G.-J.E. (2000). Behaviour of Listeria monocytogenes and autochthonous flora on meat stored under aerobic, vacuum and modified atmosphere packaging conditions with or without the presence of oregano essential oil at 5 ^{0}C. Journal of Applied Microbiology 89, 901–909.

67. Smith-Palmer, A., Stewart, J., Fyfe, L. (2001). The potential application of plant essential oils as natural food preservatives in soft cheese. Food Microbiology 18, 463– 470.

68. Shelef, L.A., Jyothi, E.K., Bulgarelli, M.A. (1984). Growth of enteropathogenic and spoilage bacteria in sage-containing broth and foods. Journal of Food Science 49 (737–740), 809.

69. Skandamis, P.N., Nychas, G.-J.E. (2001). Effect of oregano essential oil on microbiological and physico-chemical attributes of minced meat stored in air and modified atmospheres. Journal of Applied Microbiology 91, 1011 –1022.

70. Lemay, M.-J., Choquette, J., Delaquis, P.J., Garie´py, C., Rodrigue, N., Saucier, L. (2002). Antimicrobial effect of natural preservatives in a cooked and acidified chicken meat model. International Journal of Food Microbiology 78, 217– 226.

71. Ismaiel, A.A., Pierson, M.D. (1990). Effect of sodium nitrite and origanum oil on growth and toxin production of Clostridium botulinum in TYG broth and ground pork. Journal of Food Protection 53 (11), 958– 960.

72. Koutsoumanis, K., Lambropoulou, K., Nychas, G.-J.E. (1999). A predictive model for the non-thermal inactivation of Salmonella enteritidis in a food model system supplemented with a natural antimicrobial. International Journal of Food Microbiology 49, 63– 74.

73. Harpaz, S., Glatman, L., Drabkin, V., Gelman, A. (2003). Effects of herbal essential oils used to extend the shelf life of freshwaterreared Asian sea bass fish (Lates calcarifer). Journal of Food Protection 66 (3), 410–417.

74. Bayoumi, S. (1992). Bacteriostatic effect of some spices and their utilization in the manufacture of yogurt. Chemie, Mikrobiologie, Technologie der Lebensmittel 14, 21– 26.

75. Singh, N., Singh, R.K., Bhunia, A.K., Stroshine, R.L. (2002). Efficacy of chlorine dioxide, ozone and thyme essential oil or a sequential washing in killing Escherichia coli O157:H7 on lettuce and baby carrots. Leben smittel wissenchaften und Technologien 35, 720– 729.

76. Roller, S., Seedhar, P. (2002). Carvacrol and cinnamic acid inhibit microbial growth in fresh-cut melon and kiwifruit at 4 ^{0}C and 8 ^{0}C. Letters in Applied Microbiology 35, 390– 394.

Printed by Books on Demand GmbH, Norderstedt / Germany